A Wing

Crew 6-J-6 before a mission.

and a Prayer

The personal narrative of Ralph Freund, who flew 32 missions during World War II as a B-17 flight engineer and top turret gunner and member of the 379th Bomb Group.

Copyright © 2008, 2014 by Ralph Freund

All rights reserved.

No part of this book may be reproduced in any form or by any electronic or mechanical means including information storage and retrieval systems, without permission in writing from the author.

Printed in the United States of America

ISBN 978-1-63315-308-0 Paperback

ISBN 978-1-63315-342-4 Hardcover

April 9, 2008

I have been privileged to help Ralph tell the story of his War experiences. Working with Ralph also has provided the opportunity to expand my knowledge, through his eyes, of World War II.

Carol Zuckert, Personal Historian
Remember and Record
www.rememberandrecord.com

July 4, 2014

Reprinted in color for Ralph's 90th birthday, and edited to update content.

Betsy & Michael Feinberg, Ralph's Friends

Cover Art

Boeing B-17G-45-BO Fortress 42-97229 524th BS, "Hi Ho Silver". B-17G of the 379th Bomb Group, RAF Kimbolton England 1944/45 United States National Archives from originals taken by the United States Army Air Forces.

CONTENTS

1. BEFORE THE WAR 1
2. "GREETINGS" 4
3. THE REAL WAR 7
   Flight Engineer's position and duties on
     the B-17 Flying Fortress 14
4. THE FLYING FORTRESS, THE B-17 20
   The Boeing B-17 Flying Fortress 22
5. ENCOUNTER WITH THE BUZZ BOMB 24
6. D-DAY AND THE REST OF THE MISSIONS 26
   Normandy Invasion, June 1944 29
   History of the 379th Bomb Group 33
7. AFTER THE 32ND MISSION 37
AFTERWORD 40
APPENDIX 43
   Service orders given to Raiph during his time
     in the Army Air Corp 44
   Selective Service Notification of
     Classification Card 47
   Letter from Bob Jacobson to Ralph 48
   Separation Qualification Record
     "Discharge Papers" 49
   Photographs 50
   Mission Logs 52
   Crew Members 67

1

BEFORE THE WAR

I was born on the Fourth of July, 1924 at 122 Broome Street on the lower eastside of Manhattan. We lived in a third floor walk-up apartment with a pot-belly stove heated by coal that we bought by the bucketful, one bucket of coal at a time. Three or four families shared the bathroom, but we did have our own kitchen. We were typical of newly immigrated families and did not have any money to spare.

Both of my parents were born in Austria. My momma was born Sarah Feingold in 1896 and came to the United States through Ellis Island in 1910 at age 15 or 16. My pop, Morris, was born in 1895. As was common in the early part of the twentieth century, a relative, in this case, my mother's great uncle, brought everyone over from Austria. Two of Momma's brothers ended up in New York; another brother went to Canada. My father, of course, ended up in Manhattan. To support the family, Pop drove a taxi, while Momma stayed at home. I was the fourth of five sons and had younger twin sisters. I always thought that there needed to be at least two sisters in order to balance so many brothers. My parents were Orthodox Jews, and we went to services and honored all the religious holidays.

When my parents met, Pop was a foreman in a shirt factory. He hired Momma as a needleworker. She was a beautiful blonde but she was considered a botcher (Yiddish for someone who makes mistakes). She was paid on a piecework basis, so she worked very fast and not very accurately. He loved her anyway.

As I grew up, I was close to my uncles, aunts, and cousins who lived in the Bronx. Early in 1925, my family moved to 547 Fox Street in the Bronx and subsequently moved to three other houses on Fox Street. I attended P.S. 62, and then went to P.S. 52 for junior high. I went on to Bronx Vocational High School but decided to drop out of school at age 17 to go to work. I first worked as a machinist, and then as a tool and die maker, an advanced machinist.

Before I went into the service, I had large group of friends, close friends, and we hung around quite a bit. We went to Juvenile House to dance, and to Saint Mary's Park to hang out. We looked forward to just getting together --- no "hanky-panky" with our group! Believe me, we were all clean-cut kids.

Three of my old girlfriends, Sylvia, Nora and Adele, they're my age now, we call each other pretty regularly. We really like to talk about old times. But I haven't seen them for a long time.

High school friends.

2

"GREETINGS"

When I got my "Greetings" letter from President Roosevelt, I was expecting it; I was ready to go into service. I didn't want to be different than anyone else; I had no obligations other than to serve my country. I was drafted and inducted on March 9, 1943 and sent to Camp Upton, New Jersey. I did basic training in Miami Beach, Florida and then was sent to Sheppard Field, Texas (Wichita Falls). I began by learning engineering on B-25 and B-26 airplanes. However, I became the engineer on B-17s because I was needed to replace a shicker (Yiddish for drunkard) who never showed up for duty. My classification was 748, representing an engineer.

Afterwards, for one month, I trained to be a gunner in Panama City, Florida. Together with my crew, I went to Drew Field in Tampa, Florida. As a part of flight training, the crew flew together around Florida. We went to Atlanta in April, 1944 and picked up a brand new B-17G. My crew was thrilled at the prospect of flying the new B-17G model. It had a chin turret with two .50 caliber machine guns in the nose. We then flew our new B-17 to Fort Dix, New Jersey.

Boeing B-17G Flying Fortress "Shoo Shoo Shoo Baby" at the National Museum of the United States Air Force.

The B-17G was introduced onto the Fortress production line in July of 1943, and was destined to be produced in larger numbers than any other Fortress variant.

The most readily-noticeable innovation introduced by the B-17G was the power-operated Bendix turret mounted in a chin-type installation underneath the nose. This turret was equipped with two 0.50-inch machine guns.

www.joebaugher.com/usaf_bombers/b17_16.html

While we were in Fort Dix, my parents and future wife Belle Cohen visited to see me off. Belle and I had met in the Catskill Mountains of New York when I was 17 and she was 16. Our families were spending the summer in the Catskills in a bungalow community, often known as a kochalein (Yiddish for "cook alone" where each family had their own cooking facilities rather than eating in a restaurant-type setting). We saw each other a lot, but after the summer, I returned to my home in the Bronx, and she to Brooklyn. I had girlfriends, but no one was a steady then.

In Fort Dix, Belle asked if she could write to me, and I agreed. I told Belle that I would fly over her house between 9 and 11 in the morning when I left Fort Dix on the way overseas. The pilot, a good buddy, dropped the plane down from the usual 10,000 feet to 3,000 feet over Belle's house as a farewell gesture.

Ralph and Belle

3

THE REAL WAR

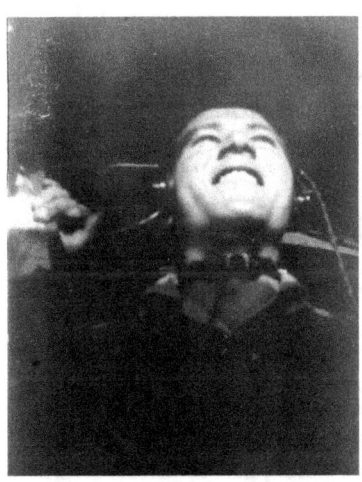

Ralph in the plane's top gunner turret.

We were stopping over in Bangor, Maine on our way to England and the war in the European Theatre of Operations. From Bangor, we went on to Newfoundland, and then at about midnight, we took off for the eleven-hour flight to Prestwick, Scotland. Cruising speed for the B-17 was usually 150 miles per hour (mph). When we were flying over the mid-Atlantic Ocean, the speed indicator dropped to 120 mph. I, the flight engineer, tried to increase the engine power to pick up speed, but the plane did not seem to respond. The indicator read that the speed had dropped to 90 mph. We were ready to throw things out of the plane to lighten the load and increase the dangerously low air speed when it dawned on me that the pitot tubes attached to the speed indicator on the wing tips were icing up. I switched on the pitot tube heaters, and to everyone's great relief, particularly mine, we were actually going 200 mph and clearly out of danger.

When we got to Scotland, the army took away our plane, much to our dismay. We really liked that plane and were disappointed. Had I not had to give up the plane, I would have named her the "Brooklyn Belle".

After the crew did more training at an area called the "Wash," we went on to the Kimbolton Airfield. The Kimbolton Airfield was located in Cambridgeshire, England, near the village of Kimbolton. It was originally built in 1941 for the Royal Air Force and expanded for American bombers' use in 1942. My crew was supposed to replace a crew that was on their 30th and last mission. We had met the crew before they took off. Unfortu-

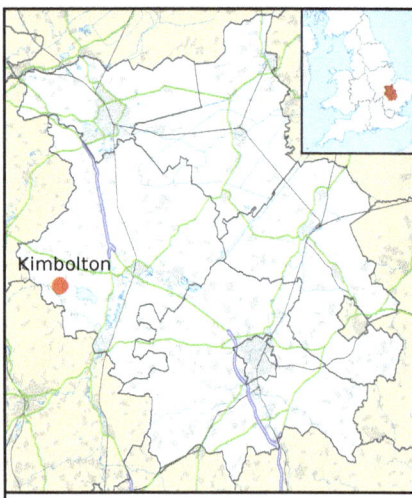

Map showing the location of RAF Kimbolton within Cambridgeshire.

nately, they never did make it back. We took over their beds, and that's the way it was. So on the next day, my crew and I began our first mission.

We completed 32 missions in a very short time span: from our first mission on May 22, 1944 to our last one, a little over two months and three days later, on July 25, 1944. We were part of the 379th Bomb Group.

The Queen Mother and Princess Elizabeth visited Kimbolton for a mock briefing. When the Queen came out on the field, all the crew stood at attention out in front of the planes so she could review the crews.

Naturally, I don't remember the specifics of each mission, but the debriefing interrogation immediately after each mission has been very useful in helping me in remembering or identifying the facts. After each mission, the whole crew contributed the details about the mission to the debriefing officer.

> Mission #1   5-22-1944 Plane: 270
> Target: Kiel German Ship yards
>
> We carried incendiaries. Forty-two bombs and 2100 gallons of fuel.
>
> The flak was moderate and inflicted minor damage. Put five holes in the wings and fuselage. Knocked the #3 turbo out by putting a hole in the exhaust.
>
> Saw one B-17 go down and saw seven men bail out. Learned later all nine men got out. Don't know yet the extent of our losses. They should be fairly light. The waist gunner saw four F.W. 190's, one of which was shot down by a P-38.
>
> The fighter protection was beautiful. The crew worked very efficiently and courageously, considering it was our first mission. Funderburg got air sick. May have to change him to another position.
>
> We took off at 8:25 returned at 14:45.
> 7:20 hr. mission.
>
> Bombing was very effective.

Editor's Note: The mission logs are in the words of the pilot, Robert Jacobson, although the entire crew attended the debriefings and contributed to them. All 32 mission logs appear in the appendix.

On the first mission we flew plane #270, a B-17 Flying Fortress. Our target was the shipyard at Kiel, Germany, about 60 miles south of the Danish border. The plane carried 42 bombs and 2100 gallons of fuel. The fighter protection was beautiful and the bombing was very effective. The crew worked very efficiently and courageously, particularly considering it was our first mission. The report shows that a B-17 got shot down and seven or possibly nine men bailed out successfully. Our plane had so many holes by the end that we couldn't count them. We, or at least I, weren't worried one bit; I was 19 years old and thought I was fearless.

However, I wasn't totally fearless. Before each mission, there were religious services. It didn't matter which religion the clergyman represented; it made no difference. We all went in to hear the clergyman pray for us and, for many, to say our own prayers. I believe in God, 100 percent, then and now. The clergyman would always recite a prayer called, "Going on a Dangerous Mission." The prayer said something like: "You are going on a dangerous mission. May God be with you. May you return safely."

Another of my beliefs, besides my strong belief in Judaism, was my yellow scarf. While we were at Fort Dix, Belle had given a me a yellow scarf, and it became one of my lucky charms. I flew every mission with the scarf. I felt that if it was good for the first mission, it would be good enough to see me through to the last mission. The crews were very superstitious, and I was no different. Some guys never changed their socks—their lucky charms. Besides Belle's scarf, another of my own superstitions was about my bicycle. I used it to ride around the base, usually from my Quonset hut to the Operations Center. I felt that if someone had stolen the bike, I would have had to quit flying because the bike was so important to me—it was more than just a good luck charm; I felt that it was saving my life; my mishegas (Yiddish for craziness).

Often, our plane went unaccompanied by fighter planes because the fighter planes could not carry enough fuel to go as far as our targets. The differences between B-17s and fighter planes were size, weight, and speed. The fighter plane could go 300 mph and had a single engine, but the B-17 had four engines and could only go 150 mph. Thankfully, after we finished 17 or 18 missions, things changed: American fighters were allowed to carry "belly tanks" under each wing. They had enough extra fuel to accompany the planes to any target and return. This was a real blessing. It cut our losses to a minimum. The fighters usually protected us very well. The belly tanks could be released when engaging the German planes so that the fighters were more maneuverable.

The crew was wonderful, very cool under fire and very compatible. Most of us flew all of the missions together. The waist gunner, Funderberg, was constantly airsick and looked green at the end of each and every mission. He was given the opportunity to leave flying duty, but chose to fly throughout the 32 missions. Our pilot was the very best, Robert (Bob) L. Jacobson. He was always calm and very pleasant.

Bob and Mary Jacobson

As the flight engineer, my position was standing behind Bob and the co-pilot, Lloyd Campbell. My job was to read and interpret the instrument panel, and relay the readings to Bob and Lloyd during take off and landing so they did not need divert their eyes. After we were in flight, I would move up to be the top turret gunner. I wielded two guns at the same time that could be rotated 360 degrees. There were a total of twelve guns on the plane. Seven of us were responsible for the twelve guns: my two, two in the ball turret, two cheek guns for the navigator, two for the waist gunners, two for the tail gunners, one for the bombardier, and a flexible .50 caliber machine gun for the radio operator.

Navigator, Raymond Balise

Bombadier, James Fulton

Radio Operator, Leonard Scarnato
　　　　Tail-Gunner, William Beard

Co-Pilot, Lloyd Campbell

Flight Engineer's position and duties on the B-17 Flying Fortress

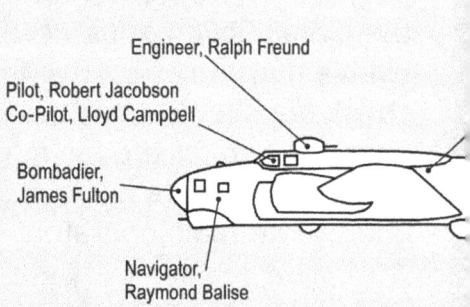

The flight engineer was specially trained to have a wide knowledge of the bomber and its equipment. He was capable of servicing the aircraft if it landed away from its home base and he could perform most jobs handled by the ground crew. Along with his ability to maintain the airframe and engines, the engineer was also an armorer with a detailed knowledge of the aircraft's guns and bomb rack. He had a working knowledge of all the aircraft systems and was a key figure in any emergency situation.

The flight engineer's primary job was manning the 8-17's top turret in combat. His view from the top turret covered a 360 degree radius over the aircraft. The turret, positioned just aft of the pilot and co-pilot on the flight deck, gave him easy access to monitor the airplane's systems. The early electric upper turret was particularly cramped with little head room but it later incorporated a higher dome with better visibility. The turret was controlled by two cycle-like hand grips. The left had the gun trigger and a safety lever. The right handle worked the range finder sight. Pulling the handles up elevated the guns and pushing them down brought them down. Pressure to the left or right rotated the turret in that direction. An interrupter stopped firing the gun if it was aimed in the propeller arc or at the tail.

At engine start and runup the flight engineer stood behind the pilots checking the fuel and engine gauges. During take-off he called off the airspeed so the pilot could concentrate on keeping the airplane straight down the runway. Once the airplane was airborne he would keep watch on the engine performance and the fuel consumption throughout the flight.

Radio man, Leonard Scarnato
Tail gunner, Asst Armorer, William Beard
Asst Engineer R.G. Funderburg
Armorer, Harry Getner
Ball turret gunner/Asst Radio man, Harry Gray

Source: www.arizonawingcaf.com

View from the tail gunner position on the B-17.

When we went out on a bombing run, the planes flew in tight formation shaped like a diamond: a lead B-17 and three following. There could be up to five diamonds—twenty planes flying together. We flew in such tight formation that we were often very close to the other planes' wings. There were close calls, times when bombs or other wings would barely miss us. Another of my responsibilities was to look up from the turret to be sure that, on a bombing run, no planes were directly above our plane. Many a time, I had to tell my pilot, "Hey Jake (Jacobson), move over" to put us out of the path of the bombs from the plane flying over us.

> Mission #2   5-23-1944 Plane #789
> Saarburcken Germany Marshaling yards
>
> We took off at 5:40 and landed at 12:20, 7:20 hrs.
>
> We carried twelve five hundred pound demolition bombs and 2100 gallons of fuel.
>
> Our main objective was an air field at Nancy-Essey. France, but the sky was 10/10 coverage. We bombed the secondary.
>
> Fighter escort was very good, but we saw no enemy fighters and flak was very weak and all of it was way off except for a little the boys reported that burst behind us.
>
> The trip was very uneventful, so there is not much to tell... Funderburg got sick again so I will try him in the top turret next trip. The crew worked very well.

Another responsibility that kept me very busy was looking out for the German fighter planes. The German fighters used the element of surprise. They came at us from 12 o'clock high and With the blindingly bright sun right in our eyes; We could hardly see them.

The thing I feared most was being shot down by German fighter planes. I also hated flak. Unfortunately, German fighter planes and flak were usually around. The bursting shells were fired from anti-aircraft guns and cannons on the ground. The anti-aircraft gunners could only estimate our altitude and then set the shells to explode at the estimated altitude. The greatest danger we encountered were enemy fighter planes.

When we returned to base, we could tell the density of flak we had encountered by damage to the plane. 10/10 refers to the heaviest of flak; 1/10 means light flak.

Our second mission was to bomb an airfield at Nancy-Essey France but cloud cover was too dense and we couldn't see out target. So we bombed the secondary target, the train tracks and supply trains at the Saarbrucken, Germany marshalling yards. We called that mission "a milk run", because it was an easy target and we were very successful.

I was happy to go on our third mission to Berlin. We knew it was German headquarters; I hoped to hit some important structures. We had no specific target identified; we were only instructed to go to the center of the city and drop our bombs. We carried 500-pound incendiary bomb clusters and 100-pound demolition bombs. The purpose of the incendiary bombs was to burn whatever they hit. They were magnesium, so water couldn't put out their flames. If the incendiary bomb hit the roof of a five-story buildin it would burn down through to the basement.

> Mission #3   5-24-1944 Plane #628
> Berlin, Germany. Target was the city itself
>
> 9:00 hr mission.
>
> We took off at 6:35 with ten five hundred pound incendiary clusters and ten one hundred pound demolition bombs. The bombs were released way too early on the I.P. and the sky coverage was about 7/10.
>
> The flak over the target was very intense and very accurate. They tracked us effectively. My bombardier was wounded in the leg and the arm shortly after bombs away. Ray made him as comfortable as possible and as soon as I could, I went down and gave him first aid. He had a very nasty wound in the leg.
>
> All the boys were very sober and serious after seeing a number of Forts explode in the air and several spin out of the formation. We were not attacked by fighters, but the wing beside us to the right was hit right after turning off the bomb run. They dove down through the formation from head on. The ball turret gunner saw two fighters hit that formation and five Forts were knocked out. All together he saw nine Forts knocked out.

The flak was heavy. We saw B-17s exploding in the air around us. Some suffered direct hits and several spun out of formation. The ball turret gunner, Ralph Gray, saw five Fortresses, "Forts", knocked down. All together he saw nine Forts knocked out. When they spun out of control, all was lost. We had over fifty holes in our plane, but we gave 'em hell. The anti aircraft guns seemed to track us pretty effectively. As I said, I was proud to be on that mission even though it cost a lot of our soldiers' lives. I hoped we had hit Hitler.

The bombardier, Fulton, who was visibly and vulnerably positioned in the nose of the plane, was wounded badly in his leg and arm on the Berlin mission. I immediately tried to give him first aid; his wound was very bad. After his recovery, he returned for our 30th mission but was shot in the same leg and in his arm. He recovered from those wounds and stayed in the service to eventually become a colonel.

4

THE FLYING FORTRESS, THE B-17

The Flying Fortress was a likeable plane because it was trustworthy. The engines were dependable. If one didn't encounter a direct hit, the plane always came back with four engines. It had a very good gliding angle. A B-17 was known to make a "dead stick landing" (out of fuel) because the gliding angle was exceptional. It was sensitive to the touch of the controls. If we ran out of fuel, the gliding angle was so true that it would help land the plane safely. A B-17 is able to land safely with only one or two engines. It was one of the best planes. It was respected and accepted that way. We often said it was able to land "on a wing and a prayer".

The Germans shot down many of our B-17s. Then they would find the shot-down B-17s, restore them and send them up where they would pretend to be part of our formation; they were intruders. They could then radio down to their anti-aircraft guns our exact altitude. Each anti-aircraft shell is set to explode at a given altitude. When the anti-aircraft guns found out our exact altitude from the intruder B-17, they could shoot at us much more accurately.

However, each plane had a specific name for each mission. For example, on one mission my plane was called "Lone Ranger". The lead plane could call in to the U.S. planes in its formation by its specific name. We knew a plane was an intruder when it didn't respond to the lead plane's call. We had quite a few intruders. The more important our target, the more intruders there were. For example, when we were flying over the Ruhr Valley in Germany, a heavily industrialized area and an important target, we had many intruders. Ball bearings were manufactured there and they were an important component of an airplane. So, the Germans would send up more intruders since they wanted to know our exact altitude. Without the intruders) all the Germans could do was guess the altitude to estimate our elevation and send up shells.

The Boeing B-17 Flying Fortress

The Boeing B-17 Flying Fortress is a four-engine heavy bomber aircraft developed in the 1930s for the United States Army Air Corps (USAAC). Competing against Douglas and Martin for a contract to build 200 bombers, the Boeing entry outperformed both competitors and exceeded the Air Corps' expectations. Although Boeing lost the contract because the prototype crashed, the Air Corps was so impressed with Boeing's design that they ordered 13 more B-17s for further evaluation. From its introduction in 1938, the B-17 Flying Fortress evolved through numerous design advances.

The B-17 was primarily employed by the United States Army Air Forces (USAAF) in the daylight precision strategic bombing campaign of World War II against German industrial and military targets. The United States Eighth Air Force, based at many airfields in southern England, and the Fifteenth Air Force, based in Italy, complemented the RAF Bomber Command's nighttime area bombing in the Combined Bomber Offensive to help secure air superiority over the cities, factories and battlefields of Western Europe in preparation for the invasion of France in 1944. The B-17 also participated to a lesser extent in the War in the Pacific, early in World War II, where it conducted raids against Japanese shipping and airfields.

From its pre-war inception, the USAAC (later USAAF) touted the aircraft as a strategic weapon; it was a potent, high-flying, long-range bomber that was able to defend itself, and to return home despite extensive battle damage. It quickly took on mythic proportions, and widely circulated stories and photos of notable numbers and examples of B-17s surviving battle damage increased its iconic status. With a service ceiling greater than any of its Allied contemporaries, the B-17 established itself as an effective weapons system, dropping more bombs than any other U.S. aircraft in World War II. Of the 1.5 million metric tons of bombs dropped on Germany and its occupied territories by U.S. aircraft, 640,000 tons were dropped from B-17s.

As of September 2011, 13 aircraft remain airworthy, with dozens more in storage or on static display.

Source: Wikipedia

5

ENCOUNTER WITH THE BUZZ BOMB

London was only 90 miles away from the Kimbolton Airfield. There was a train we could take to go to London for our passes. Belle had a British friend, Greta Abrams, whose family owned a dress factory called Golf Lane Ltd. Greta's family sent her to the United States as an exchange student to be safe from the Blitz, a sustained bombing of London by Nazi Germany in 1940 and 1941. Belle met her at Abraham Lincoln High School in Brooklyn.

For one of my 24 hour passes, I went to meet the Abrams family. I took a train from the base to London and then a taxi to where they lived in the north part of London. On the way, the taxi driver and I heard air raid sirens. This meant that the German V1s were bombing as part of the Blitz. When we heard the sirens, the taxi stopped, and the taxi driver and I jumped under the taxi. There was an expression: "Praise the Lord and keep the engine running." When the engine on the bomb stopped, it meant that the bomb was dropping and would explode. The engine cut off, and the bomb exploded two blocks away from us. It hit an empty schoolhouse.

I never did get to see Greta.

In June 1944, the German army began the use of what would be a very unique, very deadly, and historical weapon called the V1, the "vengeance weapon." Better known to Londoners as the "Buzz Bombs" or "doodlebugs," these flying bombs made a very distinct sound as they flew overhead at low altitude, before the timing mechanisms expired, and the bomb fell to earth and exploded. The people of London learned to go about their normal business and walk the streets as these huge 25-foot long, cross-shaped, bombs flew overhead. However, once they heard the engine cut off, it was time to take immediate cover, This was because as they tipped over for the final descent, the imbalance cut off the fuel flow to the engine.
Source: fighterfactory.com

6

D-DAY AND THE REST OF THE MISSIONS

Back to the missions, Mission #5 took over ten hours, longer than most missions because the

> Mission #5 5-29-1944 10:10 hrs Plane #789
> Target: Rieke Wulf factory Krrzsinki, Poland.
>
> We took off about 8:20 this morning and got back about 6:30 this evening. It was the longest mission we have had so far. I was sweating out my fuel all day. I used auto lean from the English coast back to the field.
>
> The trip was quite easy outside of being so long. We flew position in the lead squadron of the high group. A very good position. We kept very good position.
>
> We carried ten five hundred pound demolition bombs. A good load, and we really pasted the target. We got a good look at it as we left. We encountered very little flak, but we saw about five enemy planes. I saw one spin into the ground, I also saw a B-17 ditch in the channel on the way home, think it ran out of fuel. I had vision of myself doing it all day, but I made it okay. Several of the planes had to land at the coast. I guess Poland is just too far to go.

target was far from our base. The mission was to Poland and although the report doesn't mention this, I remember staying overnight in the plane so we could take off early the next morning. It was a shuttle run, three missions in one, and we refueled somewhere along the way.

I remember D-Day, June 6, 1944. The Normandy Invasion by the Allies was beginning. The invasion was originally set for the fifth of June, but bad weather and heavy seas caused General Eisenhower to delay until the next day and that date has been popularly referred to ever since by the short title "D-Day".

Mission #9 D-Day 6-6-1944 5:35 hrs Plane #004
Target: Arromanche, France.

This target was right about the middle of the beachhead established by the ground troops. We got up at 12:30 this morning and had briefing at 01:30. We figured something pretty big was up. We were briefed and sent out to our ships. We went out about an hour early to make sure everything was okay before station time. In my case I had to change ships. I flew 004.

We took off about 5:30 in the fourth squadron, flying number four. Our group put up 57 ships and our division put up 600. That is a lot of airplanes. We landed at 10:30.

The target area was crammed with allied ships, but we didn't sight any enemy craft. We only saw one burst of flak.

We carried twelve five hundred pound bombs. Some ships had two 1000 pounders hung externally. We didn't have any external shackles.

We flew in at 13,000 ft. There was an undercast below us, so we couldn't see any result.

It was our ninth mission and our target was to bomb the middle of the beachhead in Normandy where the Germans were well fortified. The German Third Reich had built an extensive system of coastal fortifications along the western coast of Europe. I remember that it was about 5:45 in the morning when we bombed. The Allied forces were leaving their landing crafts. We could see so many landing crafts that it seemed like the soldiers could almost walk from boat to boat all the way from England to France.

The Brits and the Americans flew anything with wings that day. We carried large bombs in order to penetrate German bunkers. We were flying at 5,000 feet, which was very dangerous. We were visible but this was really a surprise attack; there was no action against us. And, at 5,000 feet, we were able to be accurate. I think we did more damage in the first mission because the skies were clear, our targets were visible, and we were flying at a pretty low elevation. We could see everything plainly when we were heading back to England after the bombing.

We got back to the base at 11:00 in the morning from what had been a surprise attack: no flak, no fighters—a milk run. However, we were told that we were not to leave the base; we were going to take off again at 2 o'clock that afternoon for another mission. On the second mission of D-Day, the Germans were already waiting for us. There was flak and German fighter planes. We lost quite a few planes. Some of our planes had two 1,000-pound bombs hanging below the wings from external shackles. It was too overcast to see any of the results of our bombings.

Normandy Invasion, June 1944. On 6 June 1944 the Western Allies landed in northern France, opening the long-awaited "Second Front" against Adolph Hitler's Germany. Though they had been fighting in mainland Italy for some nine months, the Normandy invasion was in a strategically more important region, setting the stage to drive the Germans from France and ultimately destroy the National Socialist regime.

It had been four long years since France had been overrun and the British compelled to leave continental Europe, three since Hitler had attacked the Soviet Union and two and a half since the United States had formally entered the struggle. After an often seemingly hopeless fight, beginning in late 1942 the Germans had been stopped and forced into slow retreat in eastern Europe, defeated in North Africa and confronted in Italy. U.S. and British bombers had visited ruin on the enemy's industrial cities. Allied navies had contained the German submarine threat, making possible an immense buildup of ground, sea and air power in the British Isles.

Schemes for a return to France, long in preparation, were now feasible. Detailed operation plans were in hand. Troops were well-trained, vast numbers of ships accumulated, and local German forces battered from the air. Clever deceptions had confused the enemy about just when, and especially where, the blow would fall.

Commanded by U.S. Army General Dwight D. Eisenhower, the Normandy assault phase, code-named "Neptune" (the entire operation was "Overlord"), was launched when weather reports predicted satisfactory conditions on 6 June. Hundreds of amphibious ships and craft, supported by combatant warships, crossed the English Channel behind dozens of minesweepers. They arrived off the beaches before dawn. Three divisions of paratroopers (two American, one British) had already been dropped inland. Following a brief bombardment by ships' guns, Soldiers of six divisions (three American, two British and one Canadian) stormed ashore in five main landing areas, named "Utah", "Omaha", "Gold", "Juno," and "Sword". After hard fighting, especially on "Omaha" Beach, by day's end a foothold was well established.

As German counterattacks were thwarted, the Allies poured men and materiel into France. By late July these reinforcements, and constant combat made possible a break out from the Normandy perimeter. Another landing, in

> southern France in August, facilitated that nation's liberation. With the Soviets advancing from the east, Hitler's armies were shoved, sometimes haltingly and always bloodily, back toward their homeland. The Second World War had entered its climactic phase.
>
> Source: www.history.navy.mil

In general, my plane's targets were mostly in France. Germany was much further. During briefings there was a map to show us the flak areas. Aerial reconnaissance provided the troops with a very good idea of where the anti-aircraft guns were so that we knew how to avoid them.

Before our 19th mission, we had the opportunity to go to Flak House for rest and rehabilitation. However, we were flying so many missions so quickly that we figured by the time we finished the R and R, we would be finished with our required missions. So we went straight through and as I said earlier, we finished on July 25, 1944. In my 32 missions, I flew 24 different planes, none more than three times.

Getting into formation was never easy. For example, on the 20th mission to Berlin, there was an entire squadron getting into formation at the same time. We could be five feet apart; it was another time for: "Jake, move over!" We had to avoid each other; a little wind could create a collision.

> Mission #32 7-25-1944 Plane #950
> 7:10 hrs. Target: LaChapelle, France.
>
> This mission turned out to be one of the easiest of any I ever flew. We took off with 2100 gallons of fuel and 38 one hundred pound demolition bombs. We were going to just ahead of the ground forces.
>
> The bombing was very effective. We climbed to 14,000 over the field but let down to 10,000 over our target so we could bomb below the middle cloud. After bombs away we let down to 8,000 feet and came home.
>
> I let Freund fly a little formation on the way home.
>
> YOU'VE HAD IT 32 MISSIONS

I remember the easiest of all the missions, #32: our last one. In the course of the 32 missions, as far as memory serves me, I never flew the same plane more than three times because the planes had to go in for repairs very often. There were usually many flak holes, which needed to be fixed by using riveting patches, and as I mentioned before, the number of flak holes became the best way to determine how much flak action we encountered. On this last mission, Bob, the pilot, let me fly the plane. I already knew how to fly, so once in a while he would give me the controls. I have not had the controls of a plane since that last mission.

History of the 379th Bomb Group

The 379th Bomb Group was activated November 26, 1942, at Gowen Field, Boise, Idaho. It consisted of four squadrons of B-17s, the 524th, 525th, 526th and 527th. Overseas movement began in April, and in May the 379th arrived at Kimbolton, England, AAF Station 117. Its first combat mission was the bombing of German U-boat pens at St. Nazaire, France, on May 29, 1943. Colonel Maurice A. Preston was the original commanding officer until October 10, 1944, when he became the commander of the 41st Combat Wing headquartered at Molesworth. Colonel Lewis E. Lyle then assumed command of the 379th Bomb Group until May 5, 1945, when he became commander of the 41st Combat Wing. Lt. Col. Lloyd C. Mason was then named commander of the 379th Bomb Group, and was followed by lt. Col. Horace E. Frink.

The 379th Bomb Group was one of 12 heavy Bombardment Groups in the First Bombardment Division of the United States 8th Air Force. All B-17s of every Group within the 1st Bombardment Division had a large triangle painted at the top of the vertical stabilizer. Each Group's assigned code letter was painted in the triangle. The 379th's planes were assigned the letter K, and were known as the Triangle K Group.

The 379th Bomb Group flew its first 300 missions in less time than any other heavy Bombardment Group. During all of its 330 bombing missions, it dropped 26,640 tons of bombs on enemy targets, shot down 315 enemy aircraft and lost 141 of its B-17s to enemy action.

Eighty of those 141 Fortresses were shot down between May 29, 1943, and March 31, 1944. The other 61 Fortresses were lost between April 1, 1944, and April 25, 1945. One record lists 345 Fortresses assigned to the 379th Bomb Group during World War II. It is very startling that more than 43% of those 345 Fortresses were lost to enemy fighters and anti-aircraft guns.

Information in the 8th Air Force News indicates the 379th Bomb Group lost one B-17 to enemy action for every 70 sorties flown, for a loss rate of one bomber for every 22 missions. This compares to 1 bomber lost per 30 sorties by the Group with most bad fortune, and 1 bomber lost per 230 sorties for the Group with the least bad fortune. The average loss rate for the 40 Bomb Groups was 1 bomber per 88 sorties.

The 379th led the 8th Air Force in bombing accuracy, flew more sorties than any other heavy Bomb Group and had a lower loss and abortive ratio than any unit in the 8th Air Force for an extended period of time. Some of its other accomplishments include: development of the 12-plane squadron formation and 36-plane integral Group, and use of a straight-line approach on the entire bomb run.

In May 1944 it was announced that the 379th had made an unprecedented "8th Air Force Operational Grand Slam" during the preceding month. This meant that during April the 379th was first in every phase of bombing in which Bomb Groups of the 8th Air Force were graded. The 379th Bomb Group was the only unit ever awarded the 8th Air Force Grand Slam, a very unique honor that included recognition of the following achievements:

1. Best Bombing results (greatest percent of bombs on target)
2. Greatest tonnage of bombs dropped on target
3. Largest number of aircraft attacking
4. Lowest losses of aircraft
5. Lowest abortive rate of aircraft dispatched.

The 379th received two Presidential Unit Citations for its accomplishments in combat. The Group flew its last combat mission on April 25, 1945. The 379th Bomb Group remained active for two years, seven months and 29 days. During this period approximately 6,000 personnel were assigned to the Kimbolton airfield. The Group was deactivated on July 25, 1945, at Casablanca, Morocco, Africa.

Source: www.379thbga.org/history.htm

7

AFTER THE 32ND MISSION

When we returned after the 32nd mission, I immediately went to the telegraph office and wired Belle: "Your scarf was worth $32.00." Since our V-mail was censored, this was a secret way of telling her that I had completed 32 missions—my own encryption scheme.

I hung around Kimbolton for a while, and then returned to the U.S. on the Queen Elizabeth I. I got on board with $6,000 in my pocket, winnings from crap games at the base. When I got to New York, I was broke. It was gambling money, so I didn't take the loss too seriously: easy come, easy go.

It took five or six days to cross the Atlantic because we zigzagged across. This was diversionary, the way to avoid an enemy submarine from getting our ship in their sights and lining us up with a torpedo. When we landed in England at the beginning of our tours of duty, there were 20,000 troops aboard. When we returned to the United States, only 2,000 came home with us because many were still fighting. The war was still on.

Back from 32 missions over Europe as a gunner-engineer on a B-17, Tech. Sgt. Ralph Freund is shown at his home, 655 Fox St., with his bride, the former Belle Cohen, and his mother, Mrs. Sarah Freund. The couple were married on Tuesday.

Sgt. Freund, on a 21-day furlough, has received the Distinguished Flying Cross and the Air Medal with three oak leaf clusters. Mrs. Freund has three other sons in the armed forces. They are Sgt. Bernard Freund, of the Medical Corps, serving in India; Sgt. Isidore Freund, in Australia with the Air Forces, and Pvt. Max Freund, stationed at Ft. Meyers, Fla.

Local newspaper photograph of Ralph, Belle and mom shortly after Ralph's return to the United States.

I did not know my arrival date in advance. Very coincidentally, some of my wife-to-be's cousins who worked at the USO were at Pier 59 where we docked. It was mid-August 1944, so when I took the subway home, I found that most everyone was in the Catskills. I called Belle; she knew I was on my way to the Catskills in Accord, New York. They made a big party for me. I got "loaded" and sick drinking too much of too many. different things.

Belle and I got married on September 5, 1944 and spent our weeklong honeymoon at the Nevele Hotel in the Catskills in Ellenville, New York. At the time, I hadn't been discharged yet. The Army sent Belle and me down to Miami to the Ocean Grand Hotel and then to Drew Field in Tampa to the Third Air Force headquarters. I was discharged in October 1945.

AFTERWORD
(2008)

The Enemy and the Holocaust

I hated the Germans, every one of them. I hated them even more when I found out about the Holocaust. We didn't have information on how Hitler treated his enemies. We didn't know how bad the Nazis were when we were engaging them. Stories about the Holocaust never appeared in the Stars and Stripes, the military newspaper. I learned more when I returned to the United States. By then, the Holocaust was common knowledge.

Patriotism and Pride

During the Second World War, I believed that our soldiers loved our country, everybody was wonderful, and there was no animosity. I personally agreed with this sentiment. We were patriotic and loved our government and would do anything for it. I loved every minute of the army. It took care of me; it provided me with shelter; it gave me spending money. We had and still have the most wonderful army in the world; we are the best. My political attitude comes from being part of the best military in the world with the best service experience in the world. I am proud of having been a part of the United States Army.

Ralph 2014

I enjoy connecting with my experiences in the war, the good war. I decided to tell about my experiences because I believe in my country and am happy to share my story and my pride in the United States of America, then and now, among my family and my friends. Until recently, I never told anyone my story of my World War II experiences. Currently, there is an uneasy feeling about the political situation, so I want to impart some of my patriotic ideals. I am for the Iraq War because we had had enough of the Iraqi dictatorship, and it was time to do something about it. I believe that our boys are rightfully following the orders from the President to serve their country. They serve and, sadly, some die for their country, but not in vain. I want people to understand my respect for my country and I would like to help them share this respect and loyalty.

Many years ago, after my sons saw the movie, "Saving Private Ryan", they came to me and asked why I never talked about D-Day. Well, D-Day was one part of my experiences. Although it stood out, my thoughts and feeling are about my whole war-time experience.

All and all, I have had a pretty good life in spite of many personal challenges. Now that 64 years have passed since the war and I am getting older, I'm gratified to have completed my story "A Wing and a Prayer".

though
APPENDIX

Service orders given to Raiph during his time in the Army Air Corp.

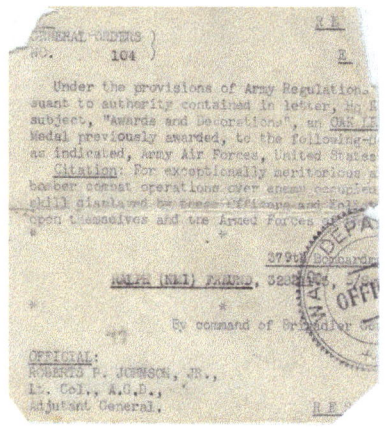

Good Conduct Medal:
For Efficiency, Honor and Fidelity

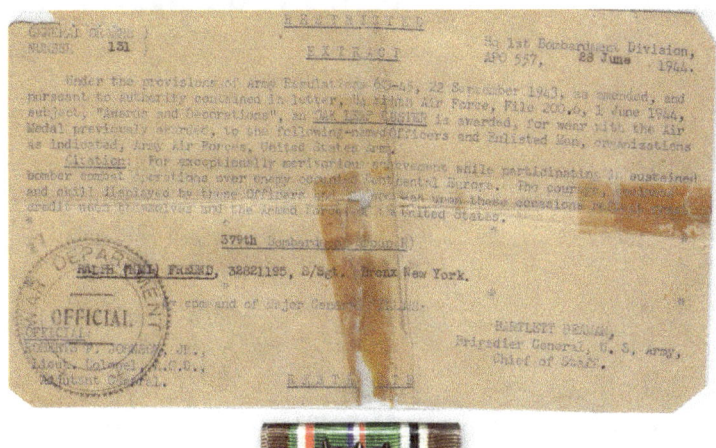

Campaign Ribbon:
"For Meritorious Achievement while Participating in Aerial Flight. The Stars represent three campaigns."

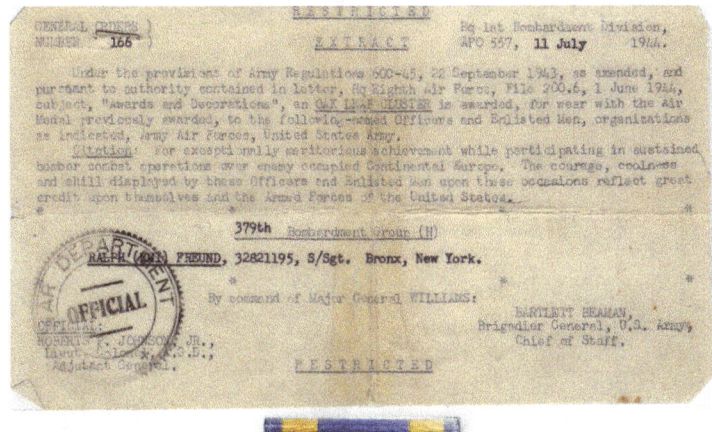

Oak Leaf Cluster:
"For exceptional meritorious achievement while participating in sustained bomber combat operations over enemy occupied Continental Europe. Each cluster represents six or seven missions."

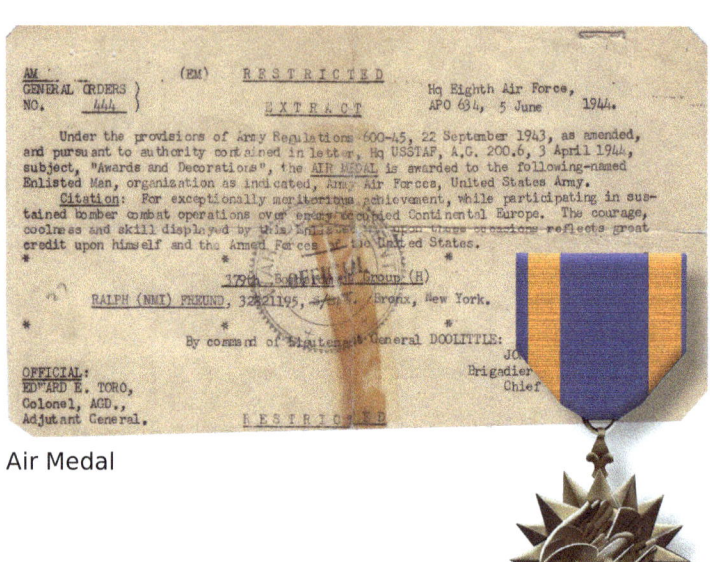

Air Medal

Distinguished Flying Cross:
"For Heroism or Extraordinary Achievement while Participating in Aerial Flight."

Selective Service Notification of Classification Card
January 1942

Letter from Bob Jacobson to Ralph
October 10, 1944

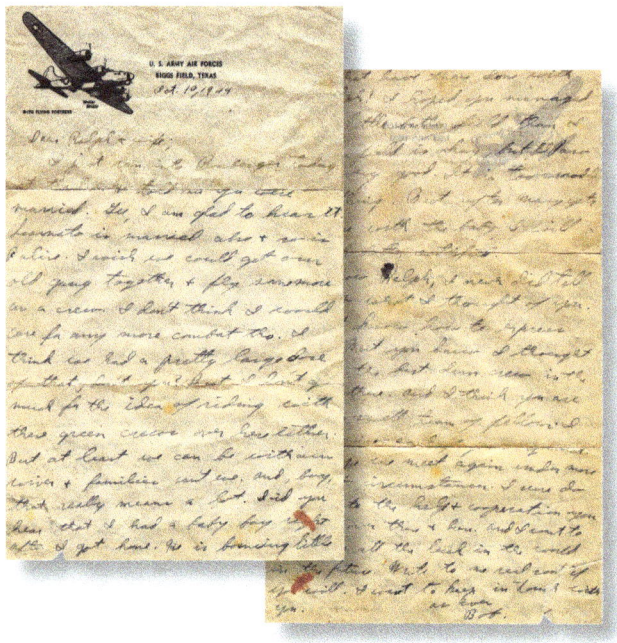

Excerpt: "...but you knew I thought you were the best darn crew in the air and I think you were a darn swell team of fellows... I hope to meet again under more pleasant circumstances. I sure do appreciate the help and cooperation you gave me over there and here and I want to wish you all the luck in the world in the future...I want to keep in touch with you. As ever, Bob" (Ralph's pilot, Bob Jacobson)

Separation Qualification Record
"Discharge Papers"
October 1945

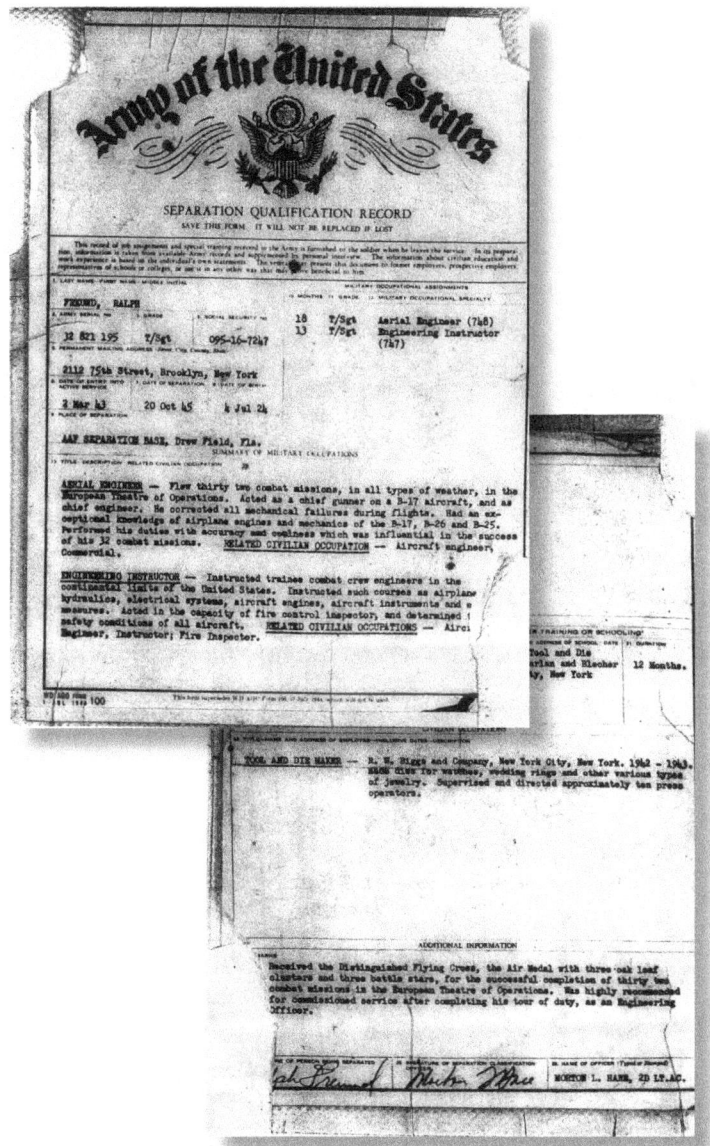

Training In Tampa Flordia

Ralph and Milton Goldstein

Ralph second from bottom

Ralph bottom left

Ralph 1944

Belle & Ralph 2000

MISSION LOGS

The following entries are the official mission logs provided during debriefings immediately following each of the 32 missions. The logs are in the words of the pilot, Robert Jacobson, although the entire crew attended the debriefings and contributed to them.

Mission #1 5-22-44 plane 270 Target Kiel, Germany. Ship yards. We carried incendiaries. Forty two bombs and 2100 gallons of fuel. The flak was moderate and inflicted minor dam- age. Put five holes in the wings and fuselage. Knocked the #3 turbo out by putting a hole in the exhaust. Saw one B-17 go down and saw seven men bail out. Learned later all nine men got out. Don't know yet the extent of our losses. They should be fairly light. The waist gunner saw four F.W. 190's, one of which was shot down by a P-38. The fighter protection was beautiful. The crew worked very efficiently and courageously, considering it was our first mission. Funderburg got air sick. May have to change him to another position. We took off at 8:25 returned at 14:45. 7:20 hr mission. Bombing was very effective.

Mission #2 5-23-44 Plane # 789 Saarburcken, Germany. Marshaling yards. We took off at 5:40 and landed at 12:20 7:20 hrs. We carried twelve five hundred pound demolition bombs and 2100 gallons of fuel. Our main objective was an air field at Nancy-Essey, France, but the sky was 10/10 coverage. We bombed the secondary. Fighter escort was very good, but we saw no enemy fighters and flak was very weak and all of it was way off except for a little the boys reported that burst behind us. The trip was very uneventful, so there is not much to tell... Funderburg got sick again so I will try him in the top turret next trip. The crew worked very well.

Mission #3 5-24-44 Plane #628 Berlin, Germany. Target was the city itself. 9:00 hr mission. We took off at 6:35 with ten five hundred pound incendiary clusters and ten one hundred pound demolition bombs. The bombs were released way too early on the I.P. and the sky coverage was about 7/10 The flak over the target was very intense and very accurate. They tracked us effectively. My bombardier was wounded in the leg and the arm shortly after bombs away. Ray made him as comfortable as possible and as soon as I could, I went down and gave him first aid. He had a very nasty wound in the leg.

All the boys were very sober and serious after seeing a number of Forts explode in the air and several spin out of the formation. We were not attacked by fighters, but the wing beside us to the right was hit right after turning off the bomb run. They dove down through the formation from head on. The ball turret gunner saw two fighters hit that formation and five Forts were knocked out. All together he saw nine Forts knocked out.

Mission #4 5-25-44 Plane #170 Sarrguemines, France. 7:20 hrs Target was marshaling yards. We took off at 5:30 and landed at 12:25. The mission was very easy and uneventful. We encountered no enemy fighters and only moderate flak over the target. We dropped ten five hundred pound demolition bombs. The bombing was very effective. Our fighter escort was beautiful to behold. I don't think we lost any ships from our group or wing. Our formation flying was very good. I had a different Bombardier today. Jim is still in the hospital.

Mission #5 5-29:44 Plane #789 Kvzsinki, Poland. 10:10 hrs Target was a Focke Wulf factory. We took off about 8:20 this morning and got back about 6:30 this evening. It was the longest mission we have had so far. I was sweating out my fuel all day. I used auto lean from the English coast back to the field. The trip was quite easy outside of being so long. We flew #3 position in the lead squadron of the high group. A very good position. We kept very good position.

We carried ten five hundred pound demolition bombs. A good load, and we really pasted the target. We got a good look at it as we left. We encountered very little flak, but we saw about five enemy planes. I saw one spin into the ground, I also saw a B-17 ditch in the channel on the way home. I think it ran out of fuel. I had vision of myself doing it all day, but I made it okay. Several of the planes had to land at the coast. I guess Poland is just too far to go.

Mission #6 5-30-44 Plane #915 Halberstadt, Germany. 7:30 hrs Factory for Ju-88 wings. We took off under instrument conditions. There was a ground fog about 500 ft. thick. Visibility was about 500 ft. We made a good take off tho. The take off was the most thrilling part of the trip. We carried ten five hundred pound bombs and 2400 gallons of fuel. We hit the target very well and encountered very little flak in the vicinity. The Navigator has been dropping our bombs ever since the Berlin raid. We earned the Air Medal on this trip. We landed late and I was so tired I couldn't even write a letter to Mary.

Mission #7 5-31-44 Plane #524 Gilze Rijen, France. 5:30 hrs Airfield. We took off at 7:50 this morning and landed about 12:30. It was planned as a much longer mission, but the clouds built up so bad we couldn't get through. So we bombed the target of last resort. We hit it very well. We carried 12 five hundred pound bombs and 2400 gallons of fuel. We used a short runway for the take off. The boys thought

I was going to take out a fence and some trees at the end of the runway, but I was holding it down to gain plenty of air speed so I wouldn't mush.

Mission #8 6-2-44 Plane #805 Paris, France. 5:45 hrs marshaling yards. We took off 5:30 in the afternoon and landed at 11 :00 that night. We carried six thousand pound bombs and 1700 gallons of fuel. We encountered no fighters, but the flak was moderate but very accurate. We got about four holes in the ship. One punctured the oil sump in #1 engine. We were lucky we didn't lose an engine. The escort was very good. All P-48s and P-5s We saw one of our escort go down from a hit by flak.

Mission #9 D-Day 6-6-44 Plane #004 Arromanche, France. 5:35 hrs This target was right about the middle of the beachhead established by the ground troops. We got up at 12:30 this morning and had briefing at 01:30. We figured something pretty big was up. We were briefed and sent out to our ships. We went out about an hour early to make sure everything was okay before station time. In my case I had to change ships. Flew 004. We took off about 5:30 in the fourth squadron, flying number four. Our group put up 57 ships and our division put up 600. That is a lot of airplanes. We landed at 10:30. The target area was crammed with allied ships, but we didn't sight any enemy craft. We only saw one burst of flak. We carried twelve five hundred pound bombs. Some ships had two 1000 pounders hung externally. We didn't have any external shackles. We flew in at 13;000 ft. There was an undercast below us, so we couldn't see any result.

Mission #10 6-7-44 Plane #462 Flers, France. 6:00 hrs Bridge head. This mission was supposed to knock out a bridge leading to the beachhead, but I am not so sure of the effectiveness of our bombing. It was done through an overcast and we two or three runs on it. We didn't get up until quite late today. We took off at 9:25 and landed about 16:00 hrs. We had to assemble above an overcast and let down through it on our return. We encountered no fighters and no flak. We bombed at 19,500 ft today. Our bomb load was twelve five hundred pounders and 1700 gallons of fuel. We lost our hydraulic pressure and it kept Lloyd busy keeping pressure up by hand, but we made it okay.

Mission #11 6-8-44 Plane #469 Orleans, France. 7:45 hrs A bridge. We bombed a bridge today and smacked the thing good. Although we had to make two runs to do it. We were loaded with six 1000 pounders and 2400 gallons of fuel. That makes a good load. We took off about 5:20 and landed at noon. We flew the #3 spot in the high squadron of the low group. We lost the turbo boost a half four before the I.P. on the #4 engine. We were able to pull back into formation and hold our position until we started on the bomb run. We dropped well behind but we caught up easily after dropping our bombs and closing the doors. We didn't see any flak to enemy fighters today. We are due to get smacked good one of these times.

Mission # 12 6-10-44 Plane #888 Vannes, France. 6:15 hrs An airfield. We got up at 12:30 this morning. I don't know what for tho. We were at our ships an hour before stations time. But we slept all afternoon, so I guess it don't much difference when we get it, just so we do get it. We took off about 05:20 this morning and landed at 10:45. We carried eighteen 250 pounders and 2100 gallons of fuel. We had very difficult flying, both over and back due to clouds and haze. Most of the ships had trouble getting into formation. We didn't see any fighters and very little flak. But we didn't drop our bombs very effectively.

Mission #13 6-12-44 Plane #462 Pontabault, France. 6:05 hrs Railroad bridge. We took off at 5:30 carrying four 2000 pounders and 1700 gallons of fuel. We had to make two runs on the target, but we hit it very well. We landed about 11:30. Lloyd took off and landed today. We only saw a little flak and no fighters. Turned 23 this day.

Mission #14 6-13-44 Plane #022 St. Andre, France. 6:20 hrs Airfield. We took off about 5:00 hrs with eighteen 250 pounders and 2400 gallons of fuel. We had enemy fighters in the vicinity, but we weren't attacked. The flak was pretty heavy, but most of it was too far away. We dropped our bombs very effectively on an airfield. We came back over England and had to hedge hop back to our base at 150 to 200 feet of altitude. The visibility was about 130 ft. It is strange several ships didn't come together. We had several close ones ourselves, but made a good landing okay and I think everybody got in.

Mission #15 6-14-44 Plane #592 Creil, France. Airfield. We took off about 4:45 today. Our load was 38 hundred pounders and 1700 gallons of fuel. Our target was an airfield near Paris. There were enemy fighters in the air, but did not attack our group. We met heavy and accurate flak over the target. As much as they threw up, I don't see how ever miss. I didn't see any ships go down over the target, but there was a lot of feathered props and stragglers. I saw one ship ditch in the channel and then blow up. The waist gunner said four chutes came out as it was going down on fire. Lloyd got bruised by a piece of flak.

Mission #16 6-16-44 Plane #022 7:30 hrs Airfield. We took off at 14:30 today and landed about 22:00 hrs. We carried twelve 500 pounders and 2100 gallons of fuel, we assembled above the overcast at 25,000 ft. It took us about a third of our fuel to assemble. We ran into a lot of flak over the target, but didn't sight any fighters. We hit the target very well. We only got a couple of flak holes.

Mission #17 6-18-44 Plane #302 7:15 hrs Hamburg, Germany. Harbor and synthetic oil plants. We took off at 5:30 and landed at 12:45. We carried eighteen 250 pounders and 2100 gallons of fuel. We had to assemble above an overcast. We hit the target very well. We saw no enemy fighters but flak was very intense and accurate. We saw one fellow get a direct hit and burst into flames. He stayed with us and dropped his bombs then left us to put the fire out He was still going down when last seen. There were numerous casualties among the ships.

Mission #18 6-19-44 Plane #003 4:50 hrs St. Omar, France. Rocket installation. We took off six this morning loaded with 38 hundred pounders and 1700 gallons of fuel. We assembled above the overcast and bombed G.H. We landed at noon. The flak was very light and we saw no bandits.

Mission #19 6-19-44 Plane #462 5:20 hrs St. Omar, France. Rocket installation. We just landed from the morning mission when they met us with a truck and took us right to lunch and back to be briefed for another mission. We took off at 14:30 and landed about 19:30. We carried twelve 500 pounders this time. It was harder, the flak was more intense and accurate. We got two quite large holes. One nearly got Ray, but we saw no bandits. On the way two ships in our group ran into each other. One of the boys claim 14 chutes got our, another says only. nine. We made out okay.

Mission #20 6-21-44 Plane #022 10:25 hrs Berlin, Germany. We took off at 4:20 this morning and assembled above a low overcast. Our load was eight 500 pound demos and two 500 pound incendiaries, plus 2700 gallons of fuel. Everything was okay until we hit the target. There we got mixed up with about four other wings and we had three hundred airplanes trying to bomb the same place at the same time. The contrails were so thick you could hardly see the element leader. We managed to be the only one of the low squadron to stay with the group. The rest climbed up above it. We went down and crossed the target indicating 200 mph or an actual airspeed of 296 mph. But we stayed with the group. We almost got scragged a couple times by other ships. They were all around us. We landed about 1400 hrs by letting down on splasher 16 through an overcast. There were fighters but our escort took very good care of them. The flak was intense but too late for our group. I imagine we lost some ships by running into each other and dropping bombs on each other.

Mission #21 6-22-44 Plane #107 7:25 hrs Lille, France. Marshaling yard. We hit the target pretty effectively. We took off at 15:30 and landed at 2100. We carried twelve 500 pounders and 1700 gallons of fuel. We didn't see any fighters. The flak was moderately heavy, but extremely accurate. We got fifteen large holes and the nose gunner and navigator both had narrow misses. One ship got his tail gunner killed, We flew low element lead.

Mission #22 6-24-44 Plane #057 7:00 hrs Bremen, Germany. Oil tanks. We took off at ten this morning and landed at 4:30. We carried eighteen 250 pounders and 2100 gallons of fuel. We flew low element lead again The flak was moderately heavy. It wasn't too accurate, but one burst hit three ships and wounded five men. Three or four seriously. One boy landed at a Limey field with four props feathered.

Mission #23 6-25-44 Plane #213 11:45 hrs Toulouse, France. Airfield. We took off at 0400 this morning carrying ten 500 pounders and 2700 gallons of fuel. General Travis led the formation. I flew lead of the low element in the low squadron. We had a hard time forming above the clouds. Nobody could seem to get together. We passed over the French coast on the way in at 15,000 feet and they shot the devil out of us. The flak was extremely thick, but not too accurate. We lost a couple ships there. I lost one of my wing men. I later lost the other one and I flew diamond most the day until Buel lost a wing man and I took his place. We made two runs on the target. When we were over the coast coming back, I talked to Butcher who was bailing out of his ship. On the way back we only had seven ships in what had been a twelve ship formation. We had to sweat our fuel. At the English coast the General said every man for himself. So not having enough fuel, we landed at Weston Zoyland, a Limey field. We refueled and took off for Kimbolton. We landed about five. The actual time in air was 11:45 hrs. Ray flew in the lead ship.

Mission #24 7-4-44 Plane 467 6:10 hrs St. Andre, France. Airfield. We took off at 3:30 with thirty eight 100 pound bombs and 2100 gallons of fuel. The lead ship got knocked out by flak, so we circled the target a couple times, all the time the flak was popping all around us. The Jerries must have know it was a big day for us to celebrate. At least the flak was different colors. Red, white and black. But no matter what color it is, I don't like it. We landed this morning with no undue damage. We could use eight more missions as easy.

Mission #25 7-5-44 Plane #888 7:40 hrs Volkel, Holland. Factory. We took off this morning with twelve 500 pounders and 1700 gallons of fuel. We couldn't bomb the primary target, so we got a secondary. The contrails were very rough, but we got very little flak. We had to sweat out our fuel. Campbell landed with only two engines and our gauges read zero. Lt. Sakryd took this plane up this afternoon after we got down and the thing blew up and killed all in the ship.

Mission #26 7-7-44 Plane #788 7:40 hrs Leipzig, Germany. Ball bearing plant. This mission was quite eventful. We took off quite early with ten 500 pounders and 2700 gallons of fuel. We assembled at altitude. Our airplane didn't have the power to hold position in the formation. We flew low diamond in the low group. We had to pull excessive power settings to even keep the formation in sight. So about ten minutes before the I.P. I blew a cylinder head on #4 engine. It didn't catch fire immediately, so I didn't feather it right then, but it soon caught fire and we had to unload our bombs near a little town.and kept turning inside of the formation to keep them in sight. But we soon lost them and we were left to ourselves. That is quite an uncomfortable feeling that close to Berlin. On the way in we saw two B-17s come together in air. It made a terrific explosion and they both went down in flames. It gives you a terrible feeling. Hannah also went down over the target, By the grace of God we made it home.

Mission #27 7-9-44 Plane #677 6:50 hrs Tours, France. Railroad bridge. We took off quite early in the morning with two 2,000 pound bombs. The trip was very uneventful. We saw only a very little flak and no fighters.

Mission #28 7-12-44 Plane #163 9:15 hrs Munich, Germany We took off fairly late this morning with four 500 pound demolition and six 500 pound incendiary bombs. The flak was very intense, but not too accurate and there were no fighters. The trip was quite uneventful except for being so long. It is pretty rough to sit still for nine hours straight. FOUR MORE TO G0!!!!!

Mission #29 7-16-44 Plane #622 9:00 hrs Munich, Germany. Munich was hit for the fourth day. We carried ten 500 pound incendiaries and 2700 gallons of fuel. The contrails over the target were terrific. I was leading the second element and had difficulty holding position in the contrails. The trip was quite uneventful except for them. The flak was light and inaccurate. Fighters were up but we didn't see any. One Me109 hit a B-17 at the back of our group at the French coast and he barely made it to the English coast on two engines. He had three men wounded. Freund (the engineer) got terribly excited in the contrails and got Lt. Wiley, who was flying co-pilot, quite nervous.

Mission #30 7-21-44 Plane #622 8:15 hrs Kaiserslautern, Germany This mission was supposed to be a D.P. which it was. But we might as well have stayed at home. We were going to bomb at 20,000 feet, but a front moved in and we had to climb to 26,000 feet to get over it We took off with 2700 gallons of fuel and 38 one hundred pound demo bombs. They are incendiaries. We didn't bomb the primary target, but picked a little town as a target of opportunity. We got a little flak after bombs away but saw no fighters. Jim rode with me this mission for the first time since the first Berlin raid.

Mission #31 7-23-44 Plane #622 6:15 hrs Creil France. This is the second time I have bombed this target. We took off late in the morning with 1700 gallons of fuel and 38 one hundred pound demolition bombs. The target was an airfield. The mission was flown beautifully as far as navigation was concerned. But we climbed to 25,000 feet over the field and circled for half an hour. We didn't have enough fuel for that. I flew the entire mission with my fuel leaned as much as possible. On the way back I decided I would leave the formation and save gas by going to the field alone. Number 4 engine quit due to lack of fuel while I was still above the overcast. So I immediately descended, and while I was still in the clouds I got #4 going again by fuel transfer. After breaking through the cloud #2 started going out. Luckily we saw an airfield and made a 180 degree turn and started an approach. We lost #2, #1 and #3 engines while still in the turn. We didn't have enough power to pull on to the field, but we touched down about 100 feet short. I bounced it good and it ballooned on up into the field. #4 cut out while we were still rolling down the runway.

Mission #32 7-25-44 Plane #950 7:10 hrs LaChapelle, France. This mission turned out to be one of the easiest of any I ever flew. We took off with 2100 gallons of fuel and 38 one hundred pound demolition bombs. We were going to just ahead of the ground forces. The bombing was very effective. We climbed to 14,000 over the field but let down to 10,000 over our target so we could bomb below the middle cloud. After bombs away we let down to 8,000 feet and came home. I let Freund fly a little formation on the way home. YOU'VE HAD IT 32 MISSIONS

CREW MEMBERS

Crew 6-J-6 Pilot Robert L. Jacobson Commissioned 8-30-43 Home Greybull, Wyo. Wife Mary E. Jacobson Address 2063-86th Ave, Oakland, Cal.

Crew 6-J-6 Co-Pilot Lloyd H. Campbell ASN 0-761560 Address 1574 Posen Ave. Berkeley, Cal. and 1030 Merced St. Berkeley, Cal. Beneficiary Ethel Ault Campbell.

Crew 6-1-6 Navigator Raymond Etalise 2nd Lt. ASN 0-707209 Address 4 Warner Rd. Leeds, Mass. Beneficiary Edward R. Raise same address.

Crew 6-J-6 Bombardier James Fulton 2nd Lt. ASN 0-762091 Address 442 N. Melbom Ave. Dearborn, Mich, Beneficiary Mary Fulton same address.

Crew 6-J-6 Engineer Ralph Freund S/Sgt, ASN 32821195 Address 655 Fox St. Bronx, New York Beneficiary Mr. J. Rapp 954 Leggett Ave Bronx, New York .

Crew 6-J-6 Asst-Engineer R.G. Funderburg ASN 18084773 Address R.R. 4 Vinita, Okla. Beneficiary Mrs. Ida J. Funderberg same address. 6-11-44 R.G. has been having considerable trouble with air sickness. I guess he will be grounded He was grounded for a few days before.

Crew 6-J-6 Radio man Leonard Scarnato S/Sgt. ASN 12134799 Address 294-8th Ave. Newark, New Jersey. Beneficiary Mrs. Rose Scarnato. He made T/Sgt.

Crew 6-J-6 Ball turret gunner-Asst radio man Harry Gray Jr. Sgt. ASN 35531762 Address RFD #6 Median, Ohio, Beneficiary Mrs. Harry Gray 2401 West Superior, Cleveland, Ohio. Made S/Sgt, 6-5-44.

Crew 6-J-6 Armorer Harry W. Getner ASN 13106123 Address 636 Rock Creek Ch. Rd. N.W. Washington, D.C, Mother Mrs. Geb. Tarlton 4301 Hafford Rd. Baltimore, Maryland, Miss Janet L. Fitzwater Elkton, Virginia. Harry was placed off flying status due to ear trouble. He was therefore the one taken off my crew when the nine man crews were formed.

Crew 6-1-6 Tail gunner-Asst Armorer. William F. Beard Sgt. ASN 20736151 Address 722 McWichita, Kan. Beneficiary Mrs. RI, Beard Same address.

www.ingramcontent.com/pod-product-compliance
Lightning Source LLC
LaVergne TN
LVHW021944060526
838200LV00042B/1923